Level 1 - ❷

Cyber Security Experts

Joy Yongo

Series Editor Casey Malarcher

Level 1 - ②

Cyber Security Experts

Joy Yongo

Series Editor: Casey Malarcher
Acquisitions Editor: Anne Taylor
Copy Editor: Liana Robinson
Cover/Interior Design: Highline Studio

ISBN: 978-1-943980-39-0

10 9 8 7 6 5 4 3 2 1
22 21 20 19 18

Photo Credits
All photos are © Shutterstock, Inc.

Contents

The Job of a Cyber Security Expert

Years ago, we had a lot of information in our wallets and houses.
It was important information.
We locked our houses and windows to keep all of this information safe.
We also kept our money in banks.

A wallet and keys ▶

Connecting to the internet

Though we still do this, many things have changed.

Much of our data is now online.

It is in cyber space.

With the internet, we can do almost anything online.

Using a bank card online ▶

A hacker

This means people can take our information.
They can find out our phone numbers, banking numbers, or even our birthdays.
They can find out our bank card numbers, passwords, and more!

Spending money online

People are not the only ones with important information online.
Businesses also have a lot of data that they don't want to share.
The government has important information, too.
All of this information could be taken by a hacker in a cyber attack.

A locked laptop

Protecting data

Cyber security experts know how to protect our data. They are the people that help keep us and the information in our computers safe.

How to Be a Cyber Security Expert

There are many ways to become a cyber security expert.
You do not even need to go to university.
But people who finish their university studies have an easier time finding jobs.

University students

Students in a computer class

You might choose to study computers at university. These days, some universities even have special classes just to teach about cyber security.

Being a cyber security expert is an important job.
It won't be your first job, though.
Instead, you will need experience.
You may work under an expert who will train you.

◀ Working with a more experienced coworker

On-the-job training ▶

A computer technician

Or you may start working on computer systems first. Maybe you will be a computer support person first. You might fix computers or keep them running well.

Security workers

Usually after years of experience, a person can work as a cyber security expert for his or her job.

Checking the security of a computer system ▶

Signing a certificate

There are also certifications for cyber security experts. Having a certification will help you become a cyber security expert faster.

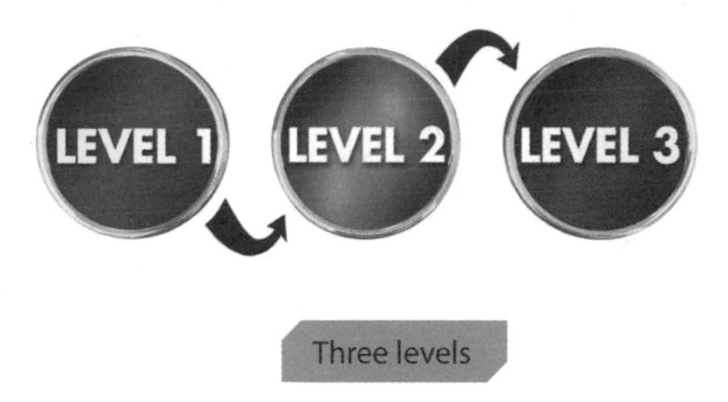

Three levels

There are usually three levels of certification.
You need to have the training and pass the test to be certified at each level.

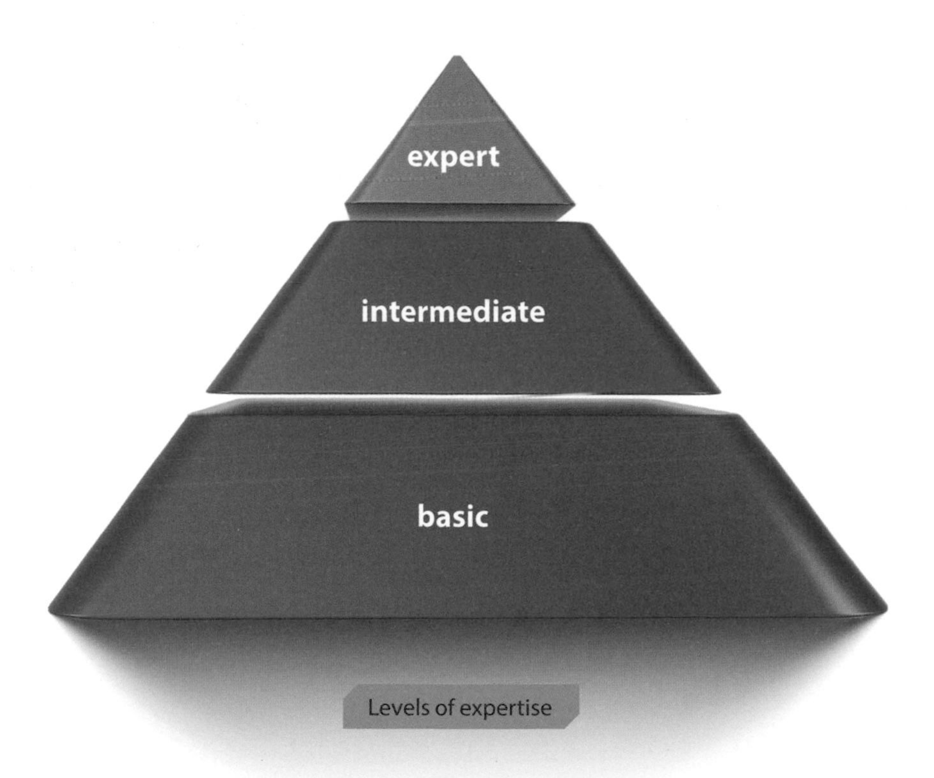

Levels of expertise

Taking an online test

◀ Answer choices

Most certifications only last from 3-5 years.
Computers and software change quickly.
You will need to take new tests to keep up with changes.

Cyber Security Experts at Work

Cyber security experts have three main jobs.
The first is to keep computer information systems safe.
This means that people who should have access to your information can see it.
And people who should not have access cannot see it.

Accessing a system

Watching over a system

To do this, cyber security experts watch over the system and its data.

The experts will also watch how computers are being used.

They are looking for anything that seems different or strange.

A security warning

Dealing with a cyber attack

As they watch over the system, cyber security experts test the security of the system.

A test shows if the security system works.

It also shows where it is not strong enough.

This is one way to stop a cyber attack before it happens.

Detecting a hacker

A cyber security expert's second job is to respond to cyber attacks.

Sometimes, a criminal wins.

He or she is able to get into a company's computer system and take data.

Hacking through the internet ▶

Responding to a cyber attack

When this happens, cyber security experts will try to find out how.

They will also check the system to see what was taken.

Security experts will work to find out where the cyber attack came from.

Then they will fix the company's security system.

They will make it stronger.

Computer code for a system

Safe data

The third part of a cyber security expert's job is teaching others.

They teach companies and people how to be safe online.

Cyber crime is different from other kinds of crime. Many times, cyber crime happens when a person clicks on the wrong link or goes to the wrong webpage. Many people do not know it has happened until it is too late.

Protecting information online

Cyber security experts can teach people what to watch out for online.
People who learn this kind of information can protect themselves.

Educating people ▶

Looking to the Future

The world is changing very quickly.
We may soon see cars that can drive themselves.
Computer parts may also be put into living things soon.
All of these new things will have software that
criminals could hack into.

A microchip ▶

A self-driving car

A cyber security expert

In the years to come, cyber security will become more and more important.

Does being a cyber security expert sound like an interesting job to you?

Comprehension Questions

1. Which of the following would not be taken in a cyber attack?
   (a) A bank card number
   (b) A birthday
   (c) A password
   (d) A microchip

2. Who do cyber security experts help?
   (a) Businesses
   (b) Governments
   (c) People like you and me
   (d) All of the above

3. According to the writer, what job might a person have before becoming a cyber security expert?
   (a) Police officer
   (b) Building security officer
   (c) Computer support person
   (d) Computer salesperson

4. Along with work experience, what might a cyber security expert need?
   (a) A fast computer
   (b) A certification
   (c) Experience as a hacker
   (d) A large team

5. Cyber security experts can prevent cyber attacks by . . .
   (a) testing the security system.
   (b) creating one password for everyone.
   (c) looking for hackers.
   (d) buying new computers.

Key **1.** (d) **2.** (d) **3.** (c) **4.** (b) **5.** (a)

Glossary

- **access** (n.) the opportunity or right to use or see something

- **certification** (n.) a document that says a person is a professional and has had special training

- **criminal** (n.) a person who breaks the law

- **cyber attack** (n. phr.) an attempt to get into or damage a computer system that is not one's own

- **hack** (v.) to secretly find a way of looking at and/or changing information on someone else's computer system without permission

- **password** (n.) a secret word or phrase that you need to know in order to be allowed into a place

- **respond** (v.) to give a spoken or written answer

- **security** (n.) the activities involved in protecting someone or something against an attack

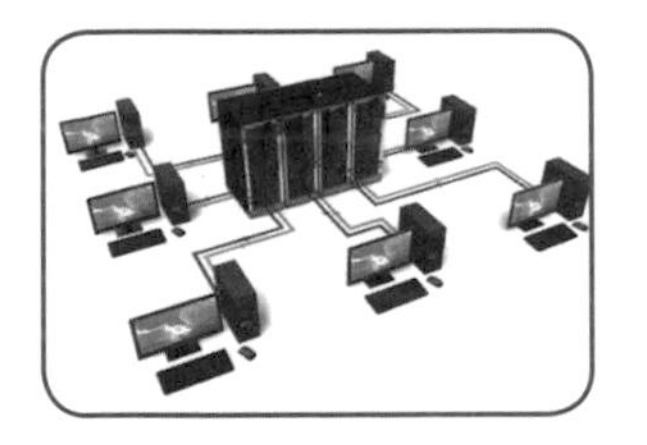

- **system** (n.) many things or parts that are put together to make a single complex whole

- **university** (n.) a place where students study after high school for an advanced degree

Notes

Here are some important cyber security software programs. Readers may enjoy researching these programs to learn more about the tools in this field.

Webroot is software that protects against viruses and malware. It prevents personal information from being stolen or captured.

Marshal software links with cloud-based services. It scans documents to see if any personal or sensitive information is being shared or can be seen readily.

SolarWinds RMM protects against malware and viruses. It offers automated monitoring, backup and recovery, and web filtering. It provides the tools needed to keep IT operations secure.

ConnectWise Automate looks for security cracks and holes. It provides both threat protection and management as well as security solutions.

List of Books

LEVEL 1

1. Robotics Engineers
2. Cyber Security Experts
3. Medical Scientists
4. Social Media Managers
5. Asset Managers

LEVEL 2

1. Drone Pilots
2. App Developers
3. Wearable Technology Creators
4. Computer Intelligence Engineers
5. Digital Modelers

LEVEL 3

1. IoT Marketing Specialists
2. Space Pilots
3. Water Harvesters
4. Genetic Counselors
5. Data Miners

LEVEL 4

1. Database Administrators
2. Nanotechnology Research Scientists
3. Quantum Computer Scientists
4. Agricultural Engineers
5. Intellectual Property Lawyers

"The future of the economy is in STEM. That's where the jobs of tomorrow will be."

James Brown (Executive Director of the STEM Education Coalition in Washington, D.C.)

Data from the US Bureau of Labor Statistics (BLS) support that assertion. Employment in occupations related to STEM—science, technology, engineering, and mathematics—is projected to grow to more than 9 million by 2022 [in the US alone]... Overall, STEM occupations are projected to grow faster than the average for all occupations.

from *STEM 101: Intro to Tomorrow's Jobs* US Bureau of Labor Statistics